I0797687

SNAKES

EYELASH VIPERS

Cody Koala

An Imprint of Pop!
popbooksonline.com

Hello! My name is Cody Koala

This book is filled with videos, puzzles, games, and more! Scan the QR codes* while you read, or visit the website below to make this book pop.

popbooksonline.com/ev

*Scanning QR codes requires a web-enabled smart device with a QR code reader app and a camera.

abdobooks.com

Published by Pop!, a division of ABDO, PO Box 398166, Minneapolis, Minnesota 55439.

Printed in the United States of America, North Mankato, Minnesota.

042025
082025

Cover Photo: Shutterstock Images
Interior Photos: AdobeStock; Alamy; Shutterstock Images; Getty Images
Editors: Elizabeth Andrews and Tyler Gieseke
Series Designers: Neil Klinepier, Colleen McLaren

Library of Congress Control Number: 2024948407

Publisher's Cataloging-in-Publication Data
Names: Murray, Julie, author.
Title: Eyelash vipers / by Julie Murray
Description: Minneapolis, Minnesota : Pop!, 2026 | Series: Snakes | Includes online resources and index
Identifiers: ISBN 9781098247812 (lib. bdg.) | ISBN 9781098248352 (ebook)
Subjects: LCSH: Vipers--Juvenile literature. | Pit vipers--Juvenile literature. | Poisonous snakes--Juvenile literature. | Snakes--Behavior--Juvenile literature. | Snakes--Juvenile literature.
Classification: DDC 597.96--dc23

Table of Contents

Chapter 1

Where Do They Live?

Eyelash vipers are **venomous** snakes. Snakes are **reptiles**. Eyelash vipers live in parts of Central and South America. They are also found in southern Mexico.

Watch a video here!

Eyelash vipers live in thick, warm forests. They are usually close to water, such as a river, lake, or stream. They spend most of their lives in trees and bushes.

Where Eyelash Vipers Live
Atlantic Ocean
Mexico
Central America
Pacific Ocean
South America
Eyelash Viper Range
N
W
E
S

Chapter 2

What Do They Look Like?

Eyelash vipers are small to medium-sized snakes. Females grow up to 2.5 feet (0.8m) long. They can weigh up to 15 pounds (6.8kg). Males are smaller.

Learn more here!

Eyelash vipers can be green, yellow, or shades of brown. Some have spots or speckles. Their bodies are covered in bumpy **scales**. The scales protect the snakes and help them grip onto tree branches.

An eyelash viper has a wide, rectangular head.

When the snake isn't using its fangs, it puts them back in its mouth.

An eyelash viper has two sharp teeth, called fangs, on the top of its mouth. **Venom** comes out of the teeth.

Eyelash vipers are not **aggressive**. But they will bite if they feel they are in danger.

Eyelash vipers have large, pointed scales above their eyes. The scales look like eyelashes. This is how the snakes got their name. These special scales help **camouflage** them in the trees.

Chapter 3

How Do They Hunt?

Eyelash vipers are **ambush** hunters. They sit and wait for their **prey**. They have pits on each side of their head. The pits sense heat. This helps vipers find prey in the dark.

Explore links here!

Eyelash vipers strike their prey at lightning speeds. They put **venom** into their prey and wait for it to die. Eyelash vipers eat lizards, frogs, mice, and birds.

Eyelash vipers attack their prey at up to 100 miles per hour (161kph)!

Chapter 4

Eyelash Viper Babies

Female eyelash vipers give birth to live babies. The eggs hatch inside the mother. She has up to 20 babies at a time. Eyelash vipers often live up to 10 years in the wild.

Baby snakes are called snakelets.

Complete an activity here!

Making Connections

Text-to-Self

Eyelash vipers spend most of their time in trees and bushes. If you were a snake, would you do the same? Why or why not?

Text-to-Text

Have you read any books about other venomous snakes? How are those snakes like eyelash vipers? How are they different?

Text-to-World

Eyelash vipers have fangs. Can you think of any other animals that have fangs?

Glossary

aggressive – mean and ready to fight.

ambush – a surprise attack made from a hidden place.

camouflage – to hide by coloring or covering to look like the surroundings.

prey – an animal that is hunted by other animals for food.

reptile – a cold-blooded animal with a skeleton inside its body and dry scales or hard plates on its skin.

scales – small, hard, thin plates that cover reptiles.

venomous – making a fluid, called venom, that is a poison to humans and animals.

Index

Online Resources

popbooksonline.com

Thanks for reading this Cody Koala book!

This book is filled with videos, puzzles, games, and more! Scan the QR codes* while you read, or visit the website below to make this book pop.

popbooksonline.com/ev

*Scanning QR codes requires a web-enabled smart device with a QR code reader app and a camera.